Our Universe

Mercury

by Margaret J. Goldstein

Lerner Publications Company • Minneapolis

Lerner Publications Company
A division of Lerner Publishing Group
241 First Avenue North
Minneapolis, MN 55401 USA

Website address: www.lernerbooks.com

Words in **bold type** are explained in a glossary on page 30.

Library of Congress Cataloging-in-Publication Data

Goldstein, Margaret J.
    Mercury / by Margaret J. Goldstein.
       p.   cm. — (Our universe)
    Includes index.
    Summary: An introduction to Mercury, describing its place in the solar system, its physical characteristics, its movement in space, and other facts about this planet.
    ISBN: 0-8225-4648-5 (lib. bdg. : alk. paper)
    1. Mercury (Planet)—Juvenile literature. [1. Mercury Planet)] I. Title. II. Series.
    QB611 .G65 2003
    523.41—dc21
                        2002000432

Manufactured in the United States of America
1 2 3 4 5 6 — JR — 08 07 06 05 04 03

People have watched the night sky for thousands of years. They have seen the Moon, stars, and planets. They named one planet Mercury. What is Mercury like?

Mercury is a small, rocky planet near the Sun. The Sun and Mercury are part of the **solar system.**

The solar system has nine planets including Mercury. All of the planets **orbit** the Sun. To orbit the Sun means to travel around it.

THE SOLAR SYSTEM

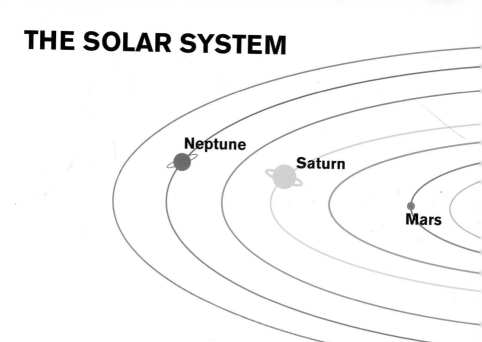

Mercury orbits in an **elliptical** path. An elliptical path is shaped like an oval. Mercury orbits close to the Sun. It is closer to the Sun than the other planets in the solar system.

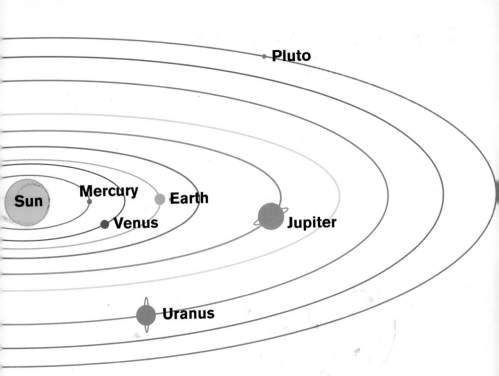

Mercury orbits faster than the other
planets. It makes one full trip around
the Sun in 88 days. Earth orbits more
slowly. Our planet travels once around
the Sun in 365 days.

Mercury also **rotates** as it travels through space. To rotate means to spin around like a top.

Mercury rotates very slowly. It spins all the way around in about 60 days. Earth takes only 1 day to rotate once.

Mercury is one of the smallest planets in the solar system. It is much smaller than Earth. Only the planet Pluto is smaller than Mercury.

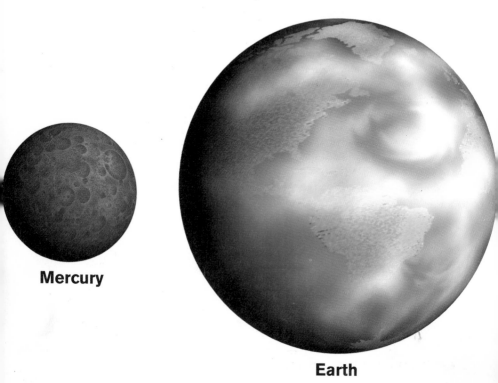

Mercury

Earth

MERCURY'S LAYERS

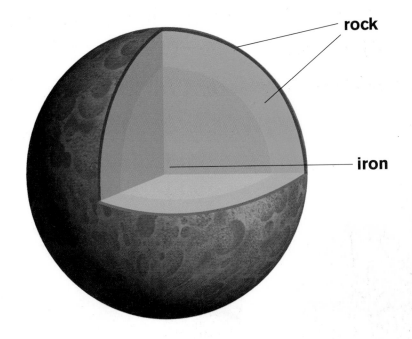

rock

iron

Mercury is made of rock and metal.
The center of the planet is made of a
metal called iron. Two layers of rock
surround the iron.

Craters cover Mercury's rocky ground. The craters look like giant holes scooped out of the planet.

The biggest crater on Mercury is called the Caloris Basin. The Caloris Basin is one of the biggest craters in the solar system. Where did Mercury's craters come from?

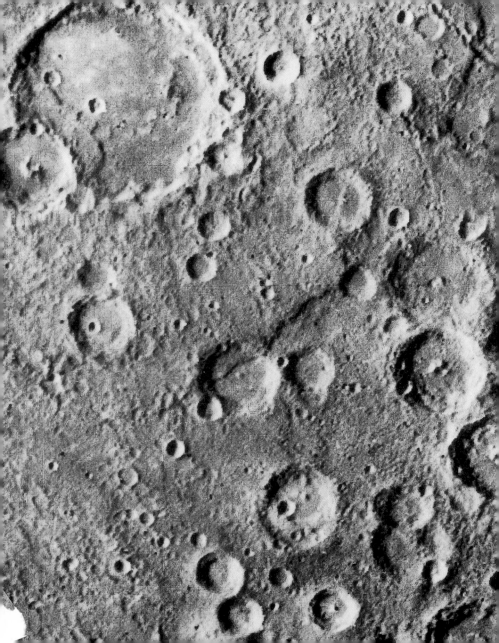

The craters were made by chunks of rock from space. They crashed into Mercury many years ago.

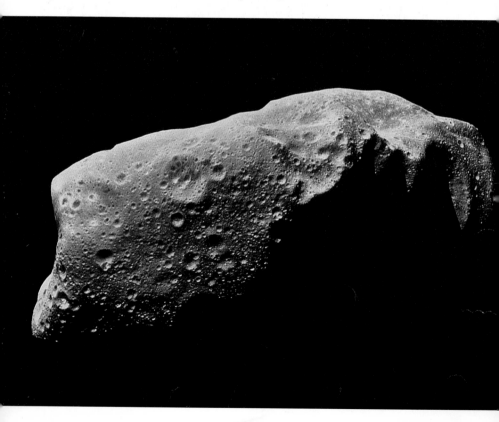

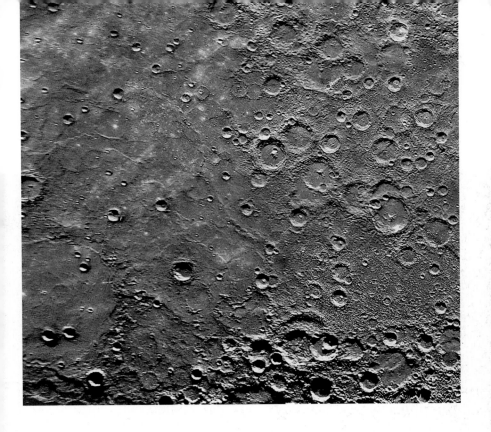

In some places on Mercury, the land
between the craters is smooth and flat.
Mercury also has tall, long cliffs that
rise up from the ground.

A thin layer of gases surrounds Mercury. This layer is called an **atmosphere.** Earth has an atmosphere, too.

The sky on Mercury is always black. It is black even during the day. No clouds float in the sky. And it never rains or snows.

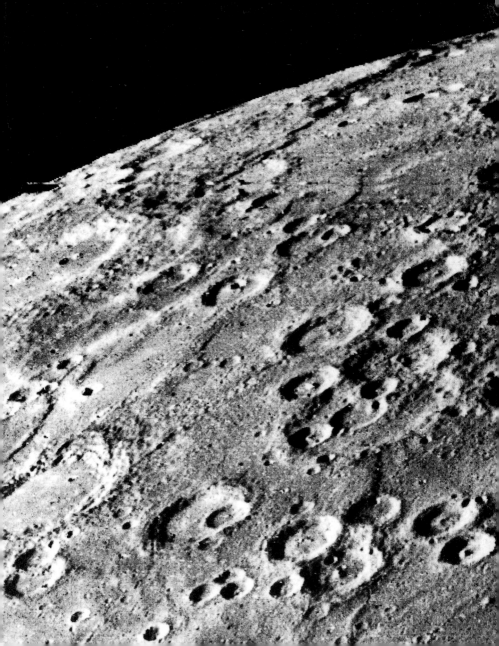

Mercury bakes in the Sun during the day. Temperatures reach hundreds of degrees above zero. That is much hotter than any place on Earth.

But Mercury is bitterly cold at night. Temperatures fall to hundreds of degrees below zero. It never gets that cold on our planet.

It is hard to see Mercury from Earth. Even powerful telescopes do not show the planet very clearly. Mercury was a mystery for many years.

In the 1970s, Americans sent a spacecraft to Mercury. The spacecraft was named *Mariner 10*. It flew near the planet in 1974 and 1975.

Mariner 10 carried cameras. The cameras took the first close-up pictures of Mercury. The pictures showed the planet's tall cliffs and many craters.

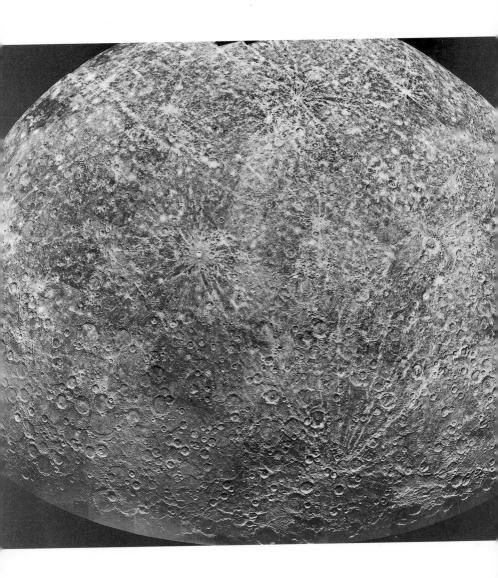

People plan to send another spacecraft to Mercury in the future. Its name is *Messenger*. *Messenger* will orbit Mercury. It will study the planet and take new pictures from space.

An artist created this picture of *Messenger.*

Imagine being in charge of *Messenger.* What would you want to know about this rocky planet near the Sun?

An artist made this picture of *Messenger* visiting Mercury.

Facts about Mercury

- Mercury is 36,000,000 miles (58,000,000 km) from the Sun.

- Mercury's diameter (distance across) is 3,030 miles (4,880 km).

- Mercury orbits the Sun in 88 days.

- Mercury rotates in 60 days.

- The average daytime temperature on Mercury is 800°F (427°C).

- The average nighttime temperature on Mercury is −300°F (−183°C).

- Mercury's atmosphere is made of helium, hydrogen, oxygen, and sodium.

- Mercury has no moons.

- Mercury was named after the messenger of the gods.

- The messenger Mercury was fast, just like the planet that carries his name.

- Mercury was visited by *Mariner 10* in 1974 and 1975.

- The only close-up photographs we have of Mercury are from *Mariner 10.* Many parts of the planet have never been seen.

- Mercury looks a lot like Earth's moon. Both Mercury and the Moon are covered with many craters.

- If you were on Mercury, the Sun would look much bigger than it does from Earth. It would seem to be more than twice as big.

- If Earth were the size of a baseball, then Mercury would be the size of a golf ball.

Glossary

atmosphere: the layer of gases that surrounds a planet or moon

crater: a large hole on a planet or moon

elliptical: shaped like an oval

orbit: to travel around a larger body in space

rotate: to spin around in space

solar system: the Sun and the planets, moons, and other objects that travel around it

Learn More about Mercury

Books

Brimmer, Larry Dane. *Mercury.* New York: Children's Press, 1999.

Simon, Seymour. *Mercury.* New York: Morrow, 1998.

Websites

Solar System Exploration: Mercury
<http://solarsystem.nasa.gov/features/planets/mercury/mercury.html>
Detailed information from the National Aeronautics and Space
Administration (NASA) about Mercury, with good links to other
helpful websites.

The Space Place
<http://spaceplace.jpl.nasa.gov>
An astronomy website for kids developed by the Jet Propulsion
Laboratory.

StarChild
<http://starchild.gsfc.nasa.gov/docs/StarChild/StarChild.html>
An online learning center for young astronomers, sponsored by
NASA.

Index